Considering the Risks of Teaching Physics More or Less Like Bean

A Whether-To Guide for Physics Faculty Big and Small

Daniel A. Martens Yaverbaum

Contents

A Tattenwitch Edition

Published by Law Four Press,

New York, NY 11213

1

Like Bean

Whoever You Are

You might

or might not wish to teach physics. If you do find yourself teaching physics or if, as turns out to be the case for many, you find yourself looking for yourself teaching physics, then it's distinctly plausible that you would prefer, in general, to be good at teaching physics — more often than not. Similarly, you might find the possibility of generally enjoying the work — deriving job satisfaction, say — on the whole more appealing than the alternative. You might be in possession of a clear set of principles or practices by virtue of which each simple act of teaching might step deliberately toward effective teaching or you might not.

If, from whatever jumble of conditions and responses, you have come to be diagnosed with an interest in teaching physics, well, then, yes: It's probably progressive. Fictional yet right statistics overwhelmingly doom your professional life to linked and looping pursuits of progress, development, self-actualization – to a drive toward self-improvement with neither kind of end particularly in any kind of sight. You are strongly advised, therefore, to shift your attention away from any dream of remedy (a fantasy) and toward the practical pursuits of maintenance and management; among the most reliable courses of proper care, argues contemporary field research literature, begins with a close reading of some teaching tips and techniques to be found in the pages within.

The contents to come are intended as public service to the physics-teaching types among us who, perhaps, are between moments of inspiration or motivation. Or possibly those who are due congratulations: three cheers for year 1 lesson 1! There is always someone somewhere pausing for reflection right before handing out a final final exam, and then there are the instructors who will receive instruction only by accident or coercion. This brief primer is designed to be of service to any of the above, but most of all to most of all: to the physics instructor so professional that he's 'just doing his job': but the one who, therefore, might turn and look this way as part of his ongoing openness – if not sprint – toward any seed or sign of some fresh challenging twist to be grafted onto some polished topic which had just recently achieved sufficient polish to want for nothing.

There might be more than one sensible model for effective physics teaching, and some models certainly might work better for some individuals than they do for others. Some might know which for whom and what for when. Such variety and complexity of approach is beyond the scope of this document. Here, we start by knowing this: Many ways to teach physics do *not* work. It is hardly a trick to know just enough physics to make a profound yet unrecognized mess of a physics class. Any way to teach physics that seems to engender both satisfaction and success is worth noting. One such way is the focus here. It works hard, it works well and it will work even better tomorrow. This model for teaching physics was developed largely by a John Jay faculty member. In order to protect ourselves from confusion and him from himself, we will refer to this very real teacher as Bean or as Professor Bean.

We will occasionally assert, in fact, that close to his full name is Professor Max Bean. In doing so, we all but eliminate any chance for hard feelings or false accusations. The name, that is, is so preposterously perfect, one almost pities the novelist so amateurish that he expects us to take his caricature seriously as protagonist. So we certainly fear no unintended sense of gossip or self-reference but by he who knows who is — to the extent implied in the paragraphs to come.

The powerful and definitive yet diminutive name *'Max Bean'* would be precisely the right way to identify a dramatic character who lacks in melodrama – or, in this case, a physics pedagogue who lives in perpetual ambivalence: as to whether he is the biggest or the smallest thing in the room. Of course, the Bean of our purpose alternately and reliably turns out to be the optimal version of each one; he frequently finds himself performing precisely to specification a magnitude he had not on that occasion elected to be. Upon such recognition – that the relative magnitude of his influence or impact (relative to the room) is not quite as he expected – our Bean is wont to conclude that the room itself is of wrong size.

Kept somewhat quiet by and among the afflicted, there is a species of physics teaching which feeds on the dream of a flawless execution (of lesson, lab, discussion, assignment, exam, etc.) and which self-perpetuates by forcing the flaw. Each time. By joining in steadfast pursuit with, through and for any reasonable goal sought by students, except that of *'mastery'*. The basic commonality we here assume, in short, is that of a rising inclination to teach physics tomorrow – from some need to fix the flaws and raise the stakes from yesterday.

Not all physics teaching looks like Bean's, but Bean's teaching is uniquely right for physics —

despite what he might want to hear. In physics, a certain kind of professor can feel authentically challenged and excited when a colleague shows him a new problem—one that involves, say, a bug and the tilted windshield of an accelerating car. He can find the problem worthy in its own right, not quickly or trivial solved, and still think it's not too hard to ask of people who have never done physics before. He can, further, legitimately protest and practically 'prove' that the feature of friction, once added to the problem's mix of issues, renders the package beyond the scope of what would be fair or legitimate or reasonable to ask of students. Such a physics professor exists, and it is a bit difficult to imagine what the Spanish class equivalent, for example, might be.

If this physics professor tends truly to push the envelope of his own type, however, he can then feel taunted or dared or disappointed by his own sensitivity toward the students. Whatever next step from which we might protect students, perhaps thinks he, assumed only the steps that came before it – the steps they knew. So? From time to time, essentially every week, this very kind of professor, still a man of his word, can wind up preparing students with such specific ferocity – yet with such adherence to the curriculum and its underlying goals – that the night arrives for writing the exam: There, witnessing as he lacks the power to stop himself, he writes precisely the once-too-hard problem but supplements the swirl with the debut of a 'gust of wind' vector!

That the seemingly novel physics phenomenon of 'wind' could first burst on the scene mid-exam is, if managed properly, something that can happen in physics.

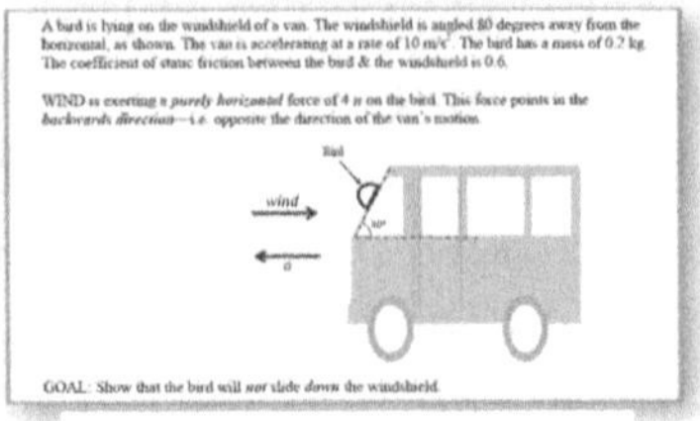

A bird is lying on the windshield of a van. The windshield is angled 80 degrees away from the horizontal, as shown. The van is accelerating at a rate of 10 m/s². The bird has a mass of 0.2 kg. The coefficient of static friction between the bird & the windshield is 0.6.

WIND is exerting a *purely horizontal* force of 4 *n* on the bird. This force points in the *backwards direction*—i.e. opposite the direction of the van's motion.

GOAL: Show that the bird will *not* slide *down* the windshield.

Figure 1: Debut of the 'Gust of Wind Vector'

That Professor Bean is the one to do it — and to manage it properly (even after all the foregoing) is, of course, the point. Bean will add the wind, so to speak, not to contrive a cheap nor intimidating surprise, but because one more force vector is just that: one more of the same and the obviously predictably natural next place physics would go. He trains himself and his students to think along such lines: the lines known as physics. Bean cannot help but write 'tricky' questions, but he could not write a 'trick' question if you paid him. Especially not then.

In physics, then, there lives a type of teaching which cultivates excellence, awe, enlightenment, and expertise, but is contemptuous of mastery. Of the anti-mastery school of schooling, Bean is the master. Education is about growth. If you seek evidence, reminder or instructive example for what growth can look like – when it is paced just slowly enough to be managed but just quickly enough to be felt – then stop reading and sit in his class... Twice. Between the two data points, you'll find a curve. That curve is a rate of change. What's changed is consciousness, and of this the students are acutely conscious, both when succeeding and struggling.

The fundamental thesis and pedagogical support offered by this document is: If you wish to pursue greatness and growth as a physics educator, then try to teach the way Bean teaches – when Bean teaches well. If you're the glutton for critique that Bean (who perhaps, as he insists, is not perfect) happens to be, then know too: Bean's weakest moments tend to be when he does not follow this advice. He does not do quite so well when he does not do quite him. If you really want to know where you might have been the least effective, ask yourself whether this too did not sound familiar.

When you feel inclined to function as a particularly powerful pedagogue, keep it up: Keep on doing what Bean would do. When you sense your grip on the group seems loose or a little less great, then stop: Right – what would Bean do? Would he stick, for example, with an abstracted instructional thread for this long – without a concrete example nor a group problem to solve? Absolutely not. Before introducing numbers, let us provide a brief overview as to Bean basics.

2

Nice and Smart, Like

The Straight Dope

To have

some competence for physics in the first place, a person has to be at least a bit of what many tend to regard as essentially 'smart', whatever that means. To have some knack for teaching, it seems reasonable to require of a candidate some evidence that s/he is 'nice'.

To be a fairly good physics teacher, then, a person had better be noticeably smart and noticeably nice – someone who teaches physics with more than basic competence, it would therefore seem, had better somehow be smart *for smart* and nice *for nice*. The pool of possibility, it seems, narrows.

With Professor Bean, the standards are met by near laughable margins. Let's explicate the basics here just in case – (and thereby practice what we preach): Regarding raw cerebral dexterity and depth, Bean is assuredly past the third standard deviation – that is, easily among the smartest people I have ever met. Regarding compassion, empathy, patience and transparent affection for persons and people alike, Bean is that: a sweetheart.* Both basic traits are clear and prominent qualities within minutes of conversation.

Of course, to be more than better – to stand out – a physics teacher then has to carve out his specific... specificity. Bean's, in most concrete terms, are:

1) He is singular at listening – actually *hearing* and *allowing* for an impact on him, by whatever newness might be sent from whomever happens to have borrowed whichever floor never demanded forever.

2) He is singular at making managing an agenda – actually progressing swiftly, clearly and responsively through a carefully packaged structure: one that opens where one might expect and keeps its own promises when it closes. When promised opening and a fulfilling close when such list is collectively called for, When Bean takes the floor, he tends to bring us to all the way to a ceiling, of whatever height and in whatever time is called for. Expansion to follow, in the form of illustration where illustrative articulation fails. It does not speak for itself, but it is going to have to – Bean hates it when I waste paper.

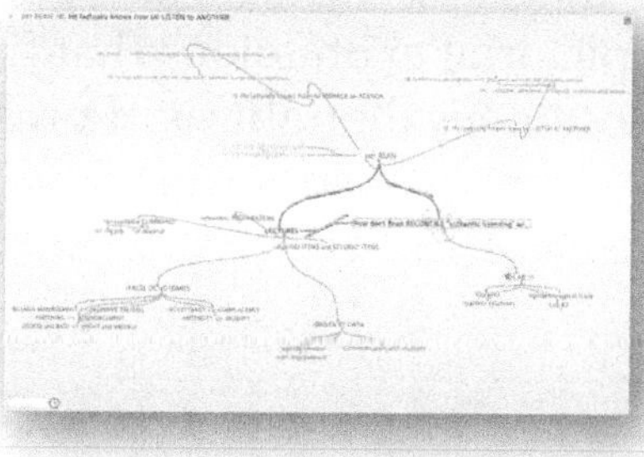

Figure 2: Some Particular yet Foggy Feedback, Valued at Less Than 1000 Words

3) While teaching is often associated with endeavors to transmit the abstract or erudite and training can be linked more to mechanical skills or the like, Bean has figured a means by which to train people into higher forms of abstract thinking. He manages this, in part, by viewing abstractions so concretely himself (as pictures, for example) that by communicating his own tidy and comfortable versions, he can get pure forms across to students with minimal intimidation or alienation (by drawing, for example). He is, to say the least, a capable draftsman. Though he might come to physics by way of math, it is the priority placed by Bean on sketches, diagrams, figures, and all imaginable pictorial expressions that firmly and repeatedly locate his thoughts and presentations within the provinces of space and time, i.e.: as belonging tof *physics*.

3

Bean

No Tricks; Technique

In one customary

format for delivering a strong and satisfying magic act — of, say, the early 20th century salon variety, a minimum of two performance threads unwind independently yet simultaneously. One such thread involves hands. Usually two. Usually those of the magician. The two hands manipulate objects (such as a deck of cards) in some kind of highly deliberate sequence of gestures. The attention of the volunteer or audience member or fine young thing or perhaps hamster, i.e.: student, is fluidly charmed, invited, dragged in, tricked out and released from a carefully constructed yet seemingly nonchalant swath of optical angles; he is thereby left with the collective impression of complete visual access and/or equivalent confidence and intimacy.

Now, then: *Independently yet simultaneously,* as our contemporary/classic Professor Bean is frequently heard to emphasize, the *'patter'* of a spoken narrative is introduced – at phrenetic transmission rates. The captivated ~~student~~ whoops-sorry the captivated *spectator* is then treated to a side of immersive lyrics to complement his hypnotic music. As long as the suite flies by with sufficient charm and/or speed, some kind of sliver of a sleight-of-hand needle – to now tracked with unblinking ferocity for just this reason – will slip its incognito way into a nether stack and the audience – hurrah! – will rejoice in another brilliant management of its own defeat: once again, the audience will go home with precisely the experience for which it planned and paid. Strangely enough, however, that experience is: *surprise.*

So where, in all the above, is Bean? Look toward the dead center – and then just stand on your head… or close your eyes… or somehow otherwise consider the powerful and precise negation of whatever you see, to wit:

Bean is a magician. Each lecture period moves as rapidly through mathematical expressions as through abstract concepts. It invariably delivers and crescendos with a surprise – something unknown built smoothly from the known. But, it hardly need be said, Bean does everything possible to keep his people with him – to look with him for the crack between moments. Whatever the students miss is at their peril and clearly not his wish. Bean sends a resounding and repeated message that the magic has taken him as well. Bean's performance is as captivating as the above, but the motivation is fundamentally more uplifting a – and more mysterious. Bean is not there to cover, but to uncover.

4

The Mysterious Stranger

DeMystified

Nothing is really obvious

. Or, perhaps some ideas make the cut (the 'obvious' cut, not just the 'true' cut) with significantly less question than some others, but I, for one, would have to think long and hard about which ideas are which.

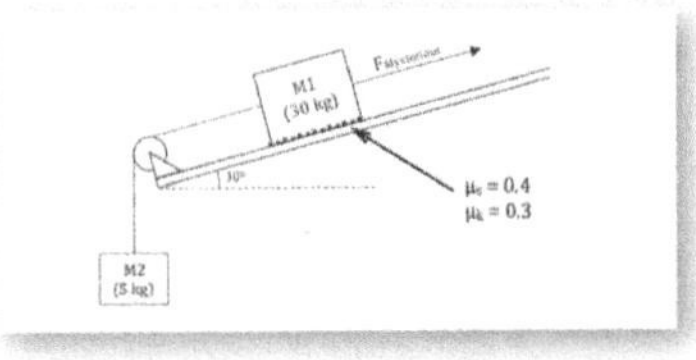

Figure 3: Fmysterious

Physics students are quite commonly heard categorizing concepts (or observations or statements, etc.) as 'obvious' (or 'easy', etc.). Sometimes the term arises in a mildly self-deprecating way, sometimes it is used to put down somebody else.

In very few cases is the term used with a conscious intent to convey a positive emotion such as excitement or esteem. That 'obvious' might be so often used in a belittling or insulting manner starts to seem less and less surprising after a while — particularly once we consider that the word is being used incorrectly and/or the correct assessment is being matched with the wrong evaluation: When learning physics for the first time, use of the term 'obvious' tends to tempt many of us immediately after the fact — just as soon as we start to detect that some mental mess of terms or numbers is settling into an identifiable shape.

When the truth of some smaller proposition reveals itself to be connected in some crucial way to a larger structure (i.e: when the fog lifts), the exhilarating floods of relief, pride and potential are, perhaps, not sufficiently captured by use of our usual agreement phrases. When we hear something we regard as true and reasonable in the moment, we can say "I agree" or "that is true" and if not, then not.

If we think something is so overwhelmingly or non-controversially true that we do not even have to think about it to know so (i.e.: 'obvious'), then we usually don't and therefore don't and neither, therefore, does the other person,... so it turns out to grow harder and harder to name an occasion to which the word 'obvious' (or 'self-evident') should properly be invited: certainly not to an exam about the five stages of cellular meiosis or to a paper on the Council of Nicea. A perfectly charming and seemingly useful word threatens to collect dust until, to consider a very frequent and invariably meaningful occurrence, Professor Bean sees a hand go up mid-lecture.

A Professor Bean lecture is, in the strictest sense, a lecture. He stands near the front where fortune has left (or the forces of pedagogical correctness have not yet found) an enormous (but insufficient) whiteboard. Students sit in rows of chairs and Bean faces them (but fully) during essentially all times he is not writing or drawing on the board.

According to at least a dozen anecdotal student reports, Bean is among a tiny minority of professors who regularly lecture without the use of Powerpoint slides or the like.

In Bean's case, 'regularly' means 'exclusively'. All sample solutions, theorem derivations, graphical renderings and verbal expressions that might be needed to support a lecture topic are made by Bean… in 'real-time'… in front of all… while his brain churns spontaneously though the (rarely trivial) physics and math needed to answer a live individual question… and while the managerial (paternal? zoological?) (group-therapeutical?) eyes in the back of his head stay trained on the collective. Bean has collected no small number of social insights, communication techniques, and professional perspectives along his background path through widely varied and distinctly challenging classroom contexts.

(Resume available upon request.) Regarding some of the phenomena or dynamics Bean did not happen to confront directly, Bean has read a lot. An impressive, intimidating lot – particularly if we take into account how respectfully and thoroughly Bean treats every word he chooses to read. The sheer number and sampling of readings that Bean has selected for mention over a couple years of curricular chats is large enough to notice. It begins to indicate a staggeringly large backdrop of research and soul-searching viz. pedagogy – once one recalls Bean's blatant disinclination to drop names or confuse authority with evidence (and once one is thereby forced to realize how much larger the unmentioned number must be).

None of the reading nor the conscious contemplation of education as a thing unto itself (never mind profession) is, so far as I can tell, visible to the students. The fruits, to be sure, are received and appreciated continually by students, but this generally seems to be in appropriately unexamined student way (thanks for that sausage, teach; I trust that you just passed it along fully made.)

There are a number of formal/institutional evaluations of Bean's teaching on file at John Jay. This means something to someone somewhere. At least, that is what I am given to understand. It's not entirely clear to me what or to whom or where, but I do feel fairly confident in stating that essentially no electronically stored information about Bean's teaching has been traded within an unregulated market nor reconstituted as elements of the cryptographic key identified with any recognized organization of international espionage. knowingly.

Similarly, I think that any official hardcopies of evaluations of Bean's teaching have been properly safeguarded from use, say, as part of any process designed, without registrar oversight, to assist other faculty members in a pursuit of better teaching or similarly suspicious activity.

There is also, however, an approximately equal number of informal/internal evaluations of Bean's teaching on file with him; he has made their contents abundantly available both to me and to the three other current members of the physics faculty. The collective student response to Bean, semester after semester, is so palpably positive that almost any word chosen to capture the specific flavor of his contribution or the specific reason for their investment invites an erroneous *under*-reading of that word at whatever scale or magnitude it is generally read to suggest.

It is quite easy to notice but quite difficult to communicate, for example, a powerful sense of ***trust*** that pervades the Bean classroom and the Bean lab. If Bean happens to be addressing an entire roster, then the students are, you can be sure, taking notes and listening. Sure—why not? Is that not what we expect in all classes? Well, yes, we might 'expect' it for a while, but sometimes the expectation sits unsatisfied until some professor finds himself volunteering for the awkwardness of asking other people to do their job and/or to act as though he is actually engaging at his.

Once students have fallen into some reasonable place of intellectual attention in many a typical non-Bean class, the para-military or perfunctory or pitiable tone with which which they are all just sucking up their academic duties and enduring another time slot is… too much to bear.

And I'm talking about a typical, not horrible, essentially decent class run by an essentially decent person. The contemporary standard by which we choose to pass most math and science classes as tolerable ways to sit and spend 75 minutes or serviceable ways to teach how to love ideas when they won't love back – is too much to bear.

The wildly disproportionate amount of time, energy and faith we coax from students toward such a disappointment is too much to bear. Our contemporary standard for the serviceable, tolerable or seemingly reasonable learning experience is too much to bear… once you've sat for a while in Bean's class (or in certain others similar in this respect). Once you've experienced what the same superficial set of physical behaviors can look like when conducted with agency by autonomous individuals who seem to feel compelled rather than convinced, inspired rather than intimidated.

So quirk in common among the questions Bean assigns, the explanations he offers, and, perhaps above all, the experimental activities he oversees, seems to prompt a certain commonality among students across differing sections and semesters. Bean's students, it eventually started to strike me, tend to talk within surprisingly similar ranges of volume and tone.

Not at all identical and with multiple exceptions, of course, but compared to expectations, that is: compared to the basic and stupefying fact that no observation like that has ever in twenty five years occurred to me even to invent about a teacher. Why would I attempt to fool even myself into that kind of crazy data? It's so absurdly unprecedented that I cannot quite even imagine what I think it's supposed to signify or why I would assume the significance to be complimentary.

But I do and it so obviously is: When a student in Bean's classroom poses a question or an insight, it is noticeably likely that the student will speak in a manner that is easily audible but is not aggressive or oppressive, that conveys investment and eagerness but does not convey too much panic or brimstone, that indicates a basic joy with whatever is going on, but seems largely free of rushes or spikes that might not be sustainable or might not be appropriate for classroom. After a short while, a great many of Bean's students, that is, start talking like him.

The resemblance, mind you, is subtle. It is not universal and I assume, of course, that it is by and large not conscious. But a very important part of the subtlety is the just that: the subtlety. Bean himself does not make his impact by being preposterously loud or flamboyantly dressed; you need not neither tolerate nor may you delight in the layers of studied-casual idiosyncrasies and self-meta-marketing ploys that have come to identify the classic professor caricature: even the seemingly low-decibel versions of the type-cast teacher have to be "whisperers". Yuck.

Not one of Bean's students, you can easily verify, would ever describe a Bean class as less than dramatic or less than stunning. They are always and memorably both those things. And no adult friend of Bean's would say that he operates entirely in the complete absence of ego or in the total liberation from vanity. Who would seek that friend? (See 'yuck', above.)

It's just that Bean finds himself teaching physics and he finds inescapable the view that a proper teaching of physics would entail the solving (by individual students, by students with one another, by students with the teacher, and by the teacher as demonstration) of many hard problems. Hard to disagree with him there. So Bean enters every lecture period with a large number of hard (quantitative, 'calculus-based') problems to solve with and in front of students.

Solving hard problems can take much time — and that's for a person who has experience solving and is solving solo. I see something of Bean on or near, I am privileged to report, a substantial percentage of his lecture days. I might be sitting in the room during lecture, I might be conferring with him right before or we might have been on the phone late on the prior night, agonizing about, for example — and for the third time in as many semesters — whether the John Jay physics program should be teaching

$$'V(ab) + V(bc) = V(ac)'$$

or

$$'V(ab) - V(ac) = V(bc)'.$$

Now, with respect simply to the bare information contained in each, it is certainly true the two statements above are substantively identical. This might or might not be obvious. As to how each expression communicates whatever we mean by such substance, as to which particular element each emphasizes, which type of sketch each motivates, which type of mathematical procedure each encourages, well, in all these respects, the statements differ enormously.

The extent to which they are both true is hardly the extent to which to they are or are not both obvious. For starters, one is phrased as a statement of addition, while the other is subtraction. Many people can do perform simple sums in their head, while many of the very same people would barely go near a minus sign with a piece of paper. That is, one of the two statements might well come across as familiar or inviting – to say nothing of right or wrong – while the other might well present an alienating abstraction. So does planning the class for a particular physics topic really begin and end with finding the true statements identified with that topic? Is it a difference that makes a difference?

To a physics teacher who feels authentically invested in the pursuit of best pedagogical practices, who wishes to discuss them in their own right, but who knows that no such discussion has a chance to start before establishing a thorough familiarity with each individual physics detail or nuance and a distractingly unscientific empathy for each real live individual student,... It is all the difference in the world. What else is there to discuss?

By his own admission, Bean might love physics – and might all be loving it even more these days, but he never loved it by itself enough to actively pursue teaching it as a profession. (Bean did not call for his job; the job called to Bean.) Of course, if Bean happened back then to have felt so strongly about physics in its own right, he would still not, presumably, have pursued the teaching of it – because the two are not one. then the pursuit of teaching it would not necessarily have been the resolving "plus a minus" vs "minus a plus" is sufficiently gripping to allow an occasional exception to the strictly self-imposed regiment of 10pm bedtime on lecture nights – if a colleague such as Yaverbaum might become available at a late hour for the phone call.

Bean's level of preparation Bean's clear mission on every lecture day is to solve problems with students – just quickly enough to cover a number of demonstrably varying examples over a number of different topics but just slowly enough both to provide annotation, connection, perspective and motivation and to not only allow but encourage student contemplation of the material. On top of that, Bean believes – and has taught me to believe – that students need room to think during class, to actually pause and perhaps convey no sign whatsoever of a progress vector, while quite possibly the greatest cognitive leap of their life might be unfolding under muzzle.

It is certainly true but might or might not be apparent (let alone obvious) that Bean, a singularly practical and grounded man in nearly all other respects, assigns to himself on a semi-weekly basis (while planning each individual lecture) and on a semi-annual basis (while planning each curriculum update proposal that I never want and always need) a mission that is plainly impossible right from the start.

True. All the above. At least six semesters prior to Bean's starting date here, I had designed a curriculum and grafted it onto the 32 (or so) class periods. It seemed to engender successful learning outcomes for many uninitiated and ignite a renewal of serious interest in some people who had seen physics in a prior stage of life. Bean took a break from his other life and visited a class period one day in 2013. For some reason, I believe it happened to be the first class day of that semester. With overt enthusiasm and approval (which meant a great deal to me), he more or less identified himself as a member of the latter camp (above).

One thing led to another and we at John Jay eventually found ourselves with the good fortune to match an open adjunct position with an available Bean. In addition to teaching the lab sections for which he was contracted, Bean spent a large and uncompensated amount of time that first semester shadowing my schedule of lectures and labs. The very first meeting took place on a Wednesday evening and entailed approx. 48 students who had to sit in one windowless room for 3 consecutive hours of fresh and unreferenced talking – there was no homework to go over, no exam to prepare for, no context of anything to fall back on.

Luckily, I had a thing or two to say, a handout or two to distribute and an exercise or two for spontaneous class consumption (which, I subsequently learned from Bean was called a "DoNow") so the inappropriately long introduction to curriculum turned out largely lively and successful. Bean seemed truly energized by the material that was presented while quite humble and open to suggestion about any prospect of ever being the one to present anything like it.

Much of those early days are now hazier to me than I care to admit, but I am fairly certain never to forget Bean's clean and compelling summary, after three hours of sweat to alchemize a fully underway class from an incoherent crowd-in-captivity, of the central lesson learned.

I had been hoping he might mirror something from the class to me, but hardly due to some interest I share with Bean in the accumulation of self-reflective data or constructive criticism. No, sir. I have hardly the skin for any of that. I was allowing myself the slight tingle of interest in feedback from Bean – only Bean – because I assumed he did not yet feel in any position to be but positive.

(That is indeed how he Bean seemed to feel. It is certainly how he shaped and delivered his numerous and specific comments whenever we were discussing an aspect of the classroom or lab or grading, etc., or when we were making explicit reference to the performance of our jobs or we were somehow reminding ourselves of our respective professional roles. Let the topic crystallize into direct discussion of physics itself, however, and anything but a full-on pursuit of truth, justice and the Newtonian way was forgotten.

My ideas were as vulnerable to scrutiny, critique and interruption as his. In such an impassioned and motivating way, every gesture of concept management is already felt so personally by Bean, that when he notices a plainly real and personal matter of someone's strength or weakness — or someone's success or failure or someone's personal experience or dream, etc., Bean steels his self against self — and handles the conversation with near Victorian elegance. Not inauthenticity, not subservience nor surrender. Just a self-confident wielding of intellect that is direct, unapologetic and warranted — a potential source of comfort even to someone who might have entered with a mind to disagree with some view of Bean's.

That part of Bean's central processing has spent countless calisthenic hours on a private prayer rug, habituating itself to juggle the most elaborately self-interlacing network of syllogistic silk and thereby exercising itself to hold steadily without squeeze onto a single separable strand — and then proceed to the next precisely and smoothly on demand. Whatever this part of the Bean thought processing process might be, it certainly does not threaten his gift for breaking down an intimidating or seemingly untethered solution into a tight sequence of tiny seemingly self-evident steps.

Performance on a solution of the kind of surgery described above is somewhat customary for a physics teacher. That piece was, I believe, quite present in a certain form in the John Jay approach prior to Bean's arrival. The ability to apply a similar knife to the construction of the *problem*, however, is a signature Bean contribution. Many of us have learned to spot the problem hidden in an exercise; to actually communicate the material of physics, it is our job to look at some scrubbed or abstracted relation among pure quantities and see the potential for a narrative to fill in the shells and for interacting elements to materialize somewhere in the arenas of space and time.

But to see within one problem precisely the right Socratic series of drills — just large enough to advance the user down a path toward the intended discovery, but just small enough so that what is f0und at the end is, in fact, a discovery, that is the art of the science.

Naturally, the above aspiration is hopelessly idealistic. Luckily, Bean is not aware that his mission is impossible. So he presses on and gets closer and closer every day to achieving that which cannot be achieved. He is, however, practical enough to know that no time is left over for him to worry about anything but classroom the curriculum the homework the grading the tests the students. He is, therefore, the raw and unpackaged substantive, Bean.

5

Why now?

Why not?

Some form

of this report ought to have been written years ago. Most reasons for the delay ought to be ignored.

One or two alleged causes for delay, while possibly ad hoc and plainly *not* exculpatory, do bear mention. They bear mention, these alleged causes for delay, because, in fact, they bear witness — to a handful of especially concrete and self- contained examples of Bean's impact. The two introductory items to follow, have faith, will tap gently toward select expository Bean character notes. The guidance evokes great thanks; we welcome any and all supporting segues into the symphony that is represented by the naturally occurring composite we are calling Professor Bean.

My first motive for not motivating has had to do with Bean's transparent desire for me to do the right thing and just write the thing do. To his own possible detriment (and as the reader would have inferred by all preceding descriptions of his character), Bean has been as polite and professional and non-imposing as possible about not at all asking ever for what was long ago rightfully his (should it only exist to be handed over). There has been no hiding, however, Bean's acute awareness that this was one responsibility of mine he could do nothing to assist, accelerate or assuage. Most infuriating to me was his pathological and insidious reason for wanting (so badly) the form:

He actually wanted — worse, expected — real feedback. That naive, thick-skinned, humble egomaniac with the last-remaining soul of a body-snatcher still dwells in some puffy-sticker village of pink ponies who can handle the truth and therefore, apparently, receive it. That sincerely deranged seeker of self-improvement (who barely has the decency to pass off his authenticity as sycophancy or ambition or something New York can stomach) really does crave and assimilate constructive criticism — in a manner that is generally so mature and rational it makes me hate his therapist.

*Note (above): I call Bean a sweetheart. All evidence tells me that his students universally agree. Some of his non-student friends, family and other relations, might describe Bean in more complex terms. They might wish to acknowledge certain respects in which Bean can occasionally be an "ass" or "bossy". No contradiction here: I have experience many flavors of Bean. I strongly believe this: Bean is allergic to asserting or imposing his *wants* over those of others – or at least anything he has identified as something he *wants*.

For a score of reasons, ranging from pragmatic to principled, Bean has no interest in elevating such things over those of others – to the point of not often identifying them. But once Bean recognizes something as a *need*, he regards such need as a priority (sometimes even emergency) to be addressed, and one that supersedes any question of 'me' or 'you' or 'smooth' or 'obnoxious' , etc. He can attack a broken idea with no concern for blood on the carpet (nor the idea's author's feelings, perhaps), but he will not acknowledge or act on some desire for credit, compensation, etc. In short, that which might be fascinating and complex in Bean's polite aggression is not so in the class. In the classroom, he is a sweetheart – who insures that the community gets what it needs.

Bean has actually been somewhat incapable of hiding his acute desire for me to write something like this and for him to read it. Because, of course, he has long since sincerely convinced himself that feedback is something he truly needs. But what he hopes to find in such a document, or what he might even — in his maddening naivete — *expect* to find in pages submitted by any professional with even a shred of basic dignity are clear and specific criticisms from which he could structure even further progress (without trying, as such) toward one of the most palpable, challenging, compassionate, uncontroversial, unconventional, sweet, smart, loving and loved educational impacts of which I have ever been aware. I have been privileged to see some truly great teachers make some truly great impacts — and often with ready and willing age groups such as high school or sixth grade. Bean, without fanfare nor intrusion, seems to changing lives in crazy ways at crazy rates — yet with college students. (!)

And he will neither shame nor shun me, when he somehow conveys with utter clarity that I failed to fulfill the requirements of the assignment – I was supposed to tell him what he's doing wrong, or could do better. With reference to specific behaviors, not judgments. With specificity not generality, etc. Oy. Well, ok, an attempt will be made. And another extension on the deadline might well be requested. So that's explanation – not excuse – number one for why this assignment is unfair: Bean has already written this paper and I'm not going to write it as well.

~

So: Whaddya say we just redistribute a bit of the challenge — or at least an aspect of the challenging feeling? Picture something like an early 6th grade round of 'truth or dare' or, if you prefer, a 'delta-epsilon proof' (regarding limits and infinitesimal quantities and often explored early in calculus classes): You name the tightest, most concrete and concise group of easily articulated Bean contributions about which you wish to hear a bit more; I guarantee that my ability to fire back a solid five pages about a man you most assuredly recognize but you most assuredly are not getting to know any better for the reading. Five pages will not get us through a title page: "Max to the min" is not an option, whether or not it comes "from the Max".

Even the little lists are too vast, elusive, meaningful and *uplifting* to abandon once begun. The website, for example, was once upon a time regarded modestly (by me) as just one of various guides through or expressions of the course material. It was a map and not necessarily the only one. It was most certainly not, therefore, the landscape.

To recognize and honor the relationship between a reality and a representation is, in fact, among the central lessons of our introductory physics course. To know that the website was an expression of a thing rather than the thing was a right recognition. To somehow half-consciously transform 'representation' into 'just a representation' was wrong; it promoted a misconception roughly related to the reasons not to spend time in biology class referring to natural selection as 'just a theory' . To tacitly permit a slow propagation of ambiguities and even errors through the web-site was, for Bean, the end of philosophical musing and the discovery of a pragmatic and increasingly urgent mess — one which demanded we sweep first, ask questions later.

Through a series of controlled and respectful movements, Prof. Bean managed to overhaul the web-site. Fully. It got to the point of reliably performing certain functions that had always been present in principle but which, in my experience, had never even pretended to work enough to allow for frustration when they did not. Students grew to take the website seriously — in the manner of a governing document. I was struck by this evolution because, if dim memories were to be trusted, that's what the website once upon a time was intended to be.

At some private moment of embarrassment, it hit me: Of course— how '*obvious*'! It might be true that the website was a 'representation of a...', rather than 'the...' but of what? Of the *curriculum.* 'Curriculum': At one time, it seemed, that word captured a fairly large and satisfying swath of the kinds of things I felt I did know something about: A curriculum, as far as I understood, involved such elements as topics, the ordering of topics and occasionally even the topic of ordering. From time to time the word could prompt a charming head-scratcher of a puzzle — such as, is topic X in my curriculum or not? If not, why not?

As I spent more and more time with Bean, however, I had begun to see that the term 'curriculum' was more dimensional than... two. or perhaps even than three. According to what I found myself gleaning from the Bean talks of early 2014, a curriculum was many-splendored; its plumage seemed sometimes more philosophical nuanced and sometimes more politically crippled than I had imagined. Whatever it was or was not, though, it was not. It was most definitively not a pile of books containing a mountain of topics, It was itself already an abstract representation of whatever it might be attempting to abstractly represent. A curriculum could be full, but it was not a filling; it was an outline or a shadow. It was, in other words and for all intents and purposes, the web-site. And now it worked. Thanks, Professor Bean!

In summary, one of Professor Bean's early and concrete contributions was the tasteful and successful war he waged on our decaying course web-site. The attack provided (for me) a nice little glimpse into a nice little Bean trait—one nice and little enough that it could, perhaps, be described. Finitely.

. . . Which will bring us back to. . . *D'oh*! the Basis for Stasis, #2. And this particular character-note/stall-tactic one, while easily as irrelevant and disingenuous as the first (described above), promises to be also convoluted and discomforting — not in some adventurously provocative way, mind you, just regrettably — like misplaced mucus. The inappropriate observation to follow, however, does turn out to be central. If an observations of any worth is ever to find itself expressed on a page sitting within an eyeshot of a Bean, then this following had better be mentioned first. We do not make measurements until we have first defined a coordinate system...

Somewhere in between the night Bean requested permission to change the "Big Ol' Duck"'s diagram and the day I saw how right he was to do so, both Bean and I began carrying some personal version of approximately the following:

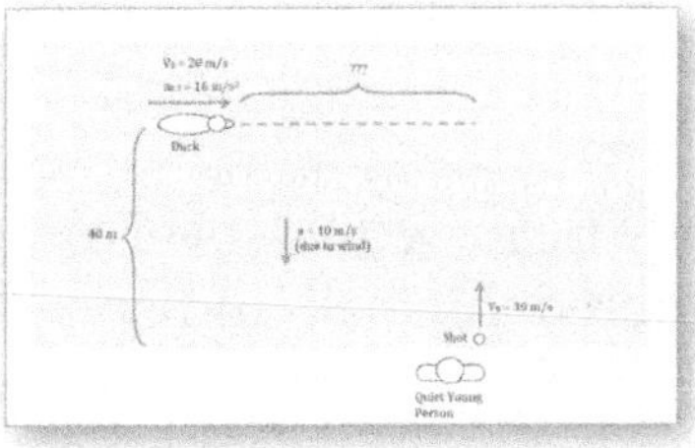

Figure 4: Bean Rehabilitates the Duck

1) At a certain point, any document of any consequence that I might write about Bean was necessarily going to include a sentence to the effect that, "Max Bean is indispensable to this physics program."

2) The above sentence was destined to be written because its meaning was destined to be true – and each of us takes teaching (or perhaps physics) too seriously either to willfully mishandle a reality as concrete as that, should one arise.

3) When you're a little kid, there comes a point of self-awareness in which you can be naked in front of the doctor and naked in front of your mom, but not both at the same time. I can tell Bean of his indispensability and I can tell John Jay for posterity (in a document), but once the witnessing is simultaneous and reciprocal, Bean will start numbering his own days. My only solace will be derive from the sheer number of numbers he knows.

Whether Prof. Bean would ever or has already literally rendered himself indispensable to the functioning of this program is of hardly any personnel/staffing consequence — compared to the sweeping and unambiguous consequence of such claim being written down. Of course, once a certain person in a certain place is identified as he without whom nothing further can be conceived, then that person has been handed keys to the certain place — and is essentially being invited to take over any door seen fit. Business schools do not tend to emphasize the rich variety of fonts and phrasings to consider when drafting and publicizing the "infinite leverage" memo.

Upon typing the obvious — that one man's feedback is another's grief certificate — I do not worry about glories he will begin to demand from the college nor corners he will begin to cut. I certainly do not worry about any impacts on his highly particular ego/esteem equilibrium. I do not worry nor wonder nor question at all, in fact. I know: that from here on in, Bean's sense of self here is doomed. Because he has, from the reference frame of record, come to comprehend the objectives and rules of our local game with mastery (brilliance) sufficient to implement a full-scale renovation — the one we forgot to order but to which we are now desperate to subscribe.

The next thing Bean needs to do here is something that, necessarily, he can only do somewhere else; right now, moreover, such is the case no matter where Bean happens to be. It's no tragedy; it's simply this (liquid-like) phase.

Conventional notions of leverage or negotiation or trajectory do not apply right now because we are in our own equilibrium: Bean's lecture instruction, his lab instruction, his curricular design, his grading, his 0% office/100% presence are all part of a perfectly balanced exchange of goods — full exploitation of Bean conducted fully on Bean terms.

That is, we at John Jay get to experience and 'count' all Bean's contributions as among our holdings; Bean, in turn, gets to *not count* this period of time (in his life) as one which he chose, as such, to devote — as such. We have all agreed —until this document started whining too loudly for attention — to cosmically configure the Bean gig in the most gloriously literal fulfillment of the noblest possible terms available for the job just sure and shabby enough to wonder in public whether it's a profession: ***calling***.

Yes, Professor Bean: You are entirely correct and not kidding yourself on this point: Everything you do here flows from the premise that we (the outside world) called you (you) and because we (the outside world) have a need and you are fulfilling it, i.e. your work here is needed and only you are truly able to do your work so in fact it was you was needed.

To repeat, you uniquely were called because you uniquely were needed. By the way, your fulfillment of the need astronomically exceeds expectations. Everybody is better than better now. We needed you to transform a place right before our very eyes and so you did. And continuously, yet. Everybody is good and you, by the way, are great. Seriously. I cannot believe that I get to say I knew you when. You, uniquely, were needed and you, uniquely, were called. Of your breathtaking day-to-day (night-to-night) contribution to this program, you are fully responsible. Of the larger semester-to-semester match made between this now and that you, however, you are fully exonerated. All understand: Choice had nothing to do with it. Your brain and being were drafted.

Of course, no place is right for your next act. Transforming a place right before our very eyes is the most we (the outside world) would ever ask for, but the pre-existing boundaries that make a place a place (in the first place) are the props and crutches for that great trick. Next is the act of no tricks. Of course, it takes place where you bring when you come. Because you'll know there (home) once you've created the zone and the system and the rules that only precisely the right hand-picked genius will be qualified to treat with a delicacy as powerful and an attack as reverential as yours. And of course everybody knows you're near-ready to do that. You've certainly had the skill and the smarts and the sensitivity and the drive for decades.

The thing that perhaps you wonder if you yet have — the thing I pray every week gets botched in delivery — is the guts to make that place than which none even remotely equivalent has been conceived. You've probably conceived it already 8.99×10^9 times. But if it's as great a zone or an axiomatic system or a region as I bet it is and if, of course, no one but you has conceived of it yet, then of course it cannot be rightly said that anyone — not one damn person — ***needs*** it, Bean.

No spelling out necessary, I'm sure. If they only find out after how incomparably great it truly is (and it truly will be), then you can't blame them (the outside world) when competition leads to commitment and commitment leads to exclusion and exclusion leaves little but the sublime to stand. I'm sorry: I will send the 99% my sincerest form of your regrets and tell them you're a bit occupied right now. But it's simply no one else's fault that you are that aware — at once of ideas and of people — that you are that smart and you are that nice.

Are you that gutsy? If, after reading among the most self-indulgent acts of self-destruction I can imagine memorializing in the wrong place at the wrong time, you still wonder, well then, I will be thrilled to see you again next semester. At 8:00 am on Fridays for lab.

Made in the USA
Monee, IL
05 November 2026

67352872R00031